Putting the Care in Customer Service

Do Your Customers Know That You Care
About Them?

Kokeita K. Miller

Edited by: Malcolm J. Lovett

Order this book online at www.trafford.com
or email orders@trafford.com

Most Trafford titles are also available at major online book retailers.

Send inquiries to: Mashun Miller, Atieko Books, 1156 Windy
Ridge Road, Chapel Hill, NC 27517-6484

Cover Design: Shamika Thomas

Printed in the United States of America.

ISBN: 978-1-4269-9227-8 (sc)
ISBN: 978-1-4269-9226-1 (e)

Trafford rev. 08/31/2011

 www.trafford.com

North America & international
toll-free: 1 888 232 4444 (USA & Canada)
phone: 250 383 6864 ♦ fax: 812 355 4082

ABOUT THE AUTHOR

The author Kokeita King Miller was born in Carrboro, North Carolina in the turbulent 1960s. She is the youngest of two children born to Eric King, deceased in 1979 and Estelle Parrish King (last name now Rayner). Her brother, Clifton L. King Sr.,deceased in 1999, always encouraged his sister to write a book. She guesses that her brother encouraged her to write because she was always so talkative.

Kokeita has held a variety of careers over the past 28 years. She has worked as a bank teller, a residence hall area director for a major university in North Carolina, a counselor for a community college in North Carolina, a workshop facilitator for the United States Air Force, a manager for a conference center, a facility manager for a community center, and currently works in the financial services industry.

In every position she has held, she has always felt that the customer, no matter their age or position in life, should be treated with the best of "CARE". Her friendly disposition and willingness to go the extra mile for the customer has brought her much success in her career. She looks at this book as a lifelong goal and hopes it will enhance the customer service skills of everyone who chooses to read it.

If you would like for Kokeita to conduct a workshop for your company or organization, please contact her at 1156 Windy Ridge Road, Chapel Hill, NC 27517 or e-mail rkmill@embarqmail.com.

To my loving family who encourages me to use all of my gifts and pursue every dream!

Randy L. Miller Sr, my husband/coach/personal trainer, Renata and Randy Jr. my children, Eric King (deceased), my dad, Estelle Parrish Rayner, my mom, Henry Rayner, my stepfather, and Clifton King (deceased), my brother

TABLE OF CONTENTS

PREFACE

"No act of kindness, no matter how small, is ever wasted." –- Aesop

One thing that we often forget is that we all have an opportunity every day to be a "customer". Whether we walk into a grocery store or call the cell phone company to complain about our bill, we are in the role of a "customer" at those times. Therefore, what I've learned while serving customers over the years is that it is best to provide outstanding customer service whenever possible. Why? Because that's the kind of customer service we all expect when we dine at restaurants, shop at retail stores, or call utility companies. After all, without customers, there is no purpose for being in business!

To stay in business--and believe me, competition is alive and well--it's all about making the customers feel like you sincerely care about their unique concern, problem, or question when they call. If you work in a call center, your main contact with the customer is over the phone. The telephone can be a very cold, impersonal way to communicate for you and the customer. The phone can be a barrier because the customer cannot see you or make an appointment with you to discuss their concerns. Furthermore, in day-to-day communication we communicate verbally and nonverbally. Without the nonverbal cues such as nodding the head, eye contact, and other body language, we are significantly limited. When we talk with customers via the telephone, we lack the use of 55% of the nonverbal cues.

This book details some special skills to help you overcome the barriers that the telephone presents. These skills will help you to establish rapport with customers quickly and communicate that you care about them, understand their problem, respect them, and appreciate their business. Many of the skills and techniques that I will discuss in this book have been developed over my 26 years in customer service positions. In addition, these skills and techniques are geared mainly for employees who work in call centers. In a call center you have to have special techniques to stay engaged in each customer's situation no matter whether they are the first customer of the day or the fiftieth customer of the day. These techniques which I have used successfully have become a natural part of my behavior. The more you practice these techniques, the more natural they become, and you may even see yourself changing the way you view the customer.

It's just like learning to ride a bike. Once you get proficient at it, you don't have to tell yourself to sit up straight on the seat and pedal. It just comes naturally. As you read this book, feel free to start with the chapters that interest you the most. Remember, the overall intention of this book is to help you become more sensitive to customer needs and learn some effective techniques for improving your rapport with them

CHAPTER ONE
Caring About The Customer

"What do we live for, if it is not to make life less difficult for each other?"—George Eliot

Illustrating that you care about the customer starts as soon as you answer the telephone. In the first few moments of the phone conversation, the tone of your voice and the words you choose should communicate care and respect for the customer. Let's look at how the term "care" is defined. The American Heritage College Dictionary defines "care" as 1) to be concerned or interested, 2) to provide needed assistance, and 3) to have a liking or attachment.

When you answer the telephone, you must immediately communicate with your greeting that you are glad that the customer called. The words that you use are important, but not nearly as important as the tone of your voice. The tone of your voice tells the customer how you are feeling when you answer the call. If your voice sounds monotone (no inflection in your voice) or flat, the caller may think the following. "Wow, I just got someone who is tired, bored, or not very interested in helping me." By the tone of your voice, customers can also detect if you are angry, happy, sad, or stressed.

The way you articulate, the tone of your voice, and the pace or speed in which you speak tells the customer a lot about you. Here are a few things that you communicate through articulation, tone, and pace: competence, knowledge, concern, sincerity, and interest in and time for their issue or question. The conclusions that we draw about people in just a small amount of time are simply amazing! The next time you call someone as a customer, pay attention to how the person answers the phone, and make a point to think about your initial feelings. Observing good and bad examples of customer service techniques can help you to improve your skills.

When you answer the telephone, you want to practice beginning with a pleasant, positive tone without sounding rushed. Invest a little time in articulating the name of your company and your name in your greeting because this will make a big difference in the customer's first impression of you. With customers of all ages calling for assistance, it is essential to pay attention to the importance of speaking clearly and articulately. Take a look below at both a good and bad example of answering the telephone.

"Thank you for calling Super Special Company, my name is Annie AskMeAnything! How can I assist you today?" (spoken in a positive, pleasant, energized tone while the associate took time to articulate each word). – Good example

"Thank you for callin Sup Spec Comp, my name is Ann Askeething. How can I assist you today?" (spoken in a hurried, uninterested tone while the words were not articulated well, and the associate sounded angry). – Bad example

If you were the customer calling in and you had a serious concern about your account, which associate would you like to speak with on the telephone? I know.

This is a no-brainer! Of course, you would want to speak to the person who sounds competent (illustrated in how words are articulated) with the energy and interest in listening to your concern and assisting you with it as well. Also, in the bad example, the person spoke in such a hurried fashion that the customer did not clearly hear every word.

Here are some tips for controlling your tone on every call. First of all, when you answer the phone, it's best to tune in right away, and put all other thoughts out of your mind. When you state your greeting, focus on presenting it in a pleasant and upbeat tone of voice. This may be challenging if you have just finished a difficult call with an angry or sad customer. However, the more you practice putting the previous call out of your mind, the easier it becomes.

Look at it this way, the next customer does not know about that call, so they should not suffer because the last customer was upset or sad. Another tip is to pretend that you are on the other end of the phone, so listen and talk to the next customer like you would want someone to talk to you.

In order to keep your tone pleasant on every call, try to find ways to keep your personal stress levels down. Overall, careers in customer service can be very selfless careers because your primary role is to help others all day long. Keep in mind that to be effective in customer service positions you must make a true effort to put the customer's needs before your own while working. This sounds difficult. But what's more rewarding than knowing that you have been there to help another person solve a problem that is causing great stress or discomfort? Something as simple as getting a letter that you don't understand from a company can be stressful for some people. Be sincere in your tone by speaking from your heart rather than just speaking empty words because people can tell the difference.

The goal for associates who answer calls is to stay positive and to greet each customer with a tone that communicates, "I care about you Mr. or Mrs. Customer and truly want to help you with your concern." It's important to communicate that you care about the customer because that makes the customer feel important and valued. A valued customer will frequently speak positively about your company to friends and business colleagues. This can result in the acquisition of new customers and increased customer retention. As your company's customer base grows, you can often benefit in these ways: 1) job stability, 2) promotions, 3) bonuses, and 4) merit increases.

As simple as it may seem, people want someone to sincerely care about their issue or question and provide an accurate answer or solution to their issue. When you speak, concentrate on the tone of your voice and speak from the sincerity of your heart.

Here are some real comments from customers who I have interacted with via telephone and in person in my role as a bank teller, career counselor, and workshop facilitator.

"I can hear the smile in your voice."

"Thank you for being so patient and kind."

"You explained the options for opening a savings and checking account with this bank in a very clear and easy to understand way."

"Your sense of humor and outgoing personality really kept my attention during the workshop today! Thank you!"

The next critical element of showing you care is to show the customer that you respect them. "R-E-S-P-E-C-T, find out what it means to me, R-E-S-P-E-C-T... All I want, ooh yeah, I want a little Respect." In the 1960s, Aretha Franklin sang about "Respect", and the song was quite popular amongst most Americans. In her song, she stressed the importance of respect, and she even taught some youngsters like me how to spell the word at a young age. In customer service positions, respect for the customer is essential in showing that you care about them and their issue. The American Heritage College Dictionary defines respect as: 1) the state of being regarded with honor or esteem, 2) willingness to show consideration or appreciation, and 3) polite expression of consideration or deference.

One of the easiest ways to communicate respect is to address customers by their last names. For example, use their name when talking with them such as "Mr. Jones" or "Ms. Jones." You want to make the best possible effort to pronounce the name correctly. (Some names are very challenging, and customers may even tell you to call them by their first names. This is permitted if you have the customer's permission.).

Addressing the customer by name helps establish rapport and respect for the customer. If you use the customer's first name, remember to keep the conversation professional.

Avoid using slang or sharing too much personal information about yourself. Focus on the customer and the reason for his/her call. The call can be conversational without getting way off track and wasting the customer's time, your company's time, and your time. Remember time equals money!

Another way to illustrate respect is by using phrases such as "yes sir" or "yes ma'am" as appropriate. When you answer a question for a customer and he/she replies, "Sorry for taking your time today," this is an excellent time to say something like, "No, thank you so much for phoning in about your concern." It's always best to communicate to customers that their call is appreciated and that the purpose of your job is to answer their inquiries and assist them with their needs. Something that I heard another associate say once seems to be received very positively by customers. When customers say thanks for taking the time to answer my question today, you can try replying with this simple yet effective phrase, "If there were no you, there would be no me." Every time I use this phrase, most customers say thank you or chuckle and then tell me I'm right.

The main point that I am stressing here is that every customer deserves the same level of respect whether he/she calls in angry or happy.

When your word choice and tone are both polite, customers know that you care about them, and in turn, 99% of them will reciprocate with respect. Keep in mind that short, snippy remarks or using an angry tone with the customer just starts a fire with most customers. Be sure to steer yourself more towards calm, polite statements of fact with a concerned, caring tone. Later in this book, you will be given some practical techniques for staying positive, even after a call from an extremely irate customer.

CHAPTER TWO
Understanding the Customer

"Happiness comes of the capacity to feel deeply, to enjoy simply, to think freely, to risk life, to be needed." – Storm Jameson

To truly understand the customer, you need to use techniques such as active listening and empathy. Active listening involves focusing on what the customers are saying as they describe their issues or questions. It is always best to take notes while they are talking. This serves two purposes: it helps you resist the urge to interrupt the customers, and it helps you to be able to accurately paraphrase what their needs are before you take action. Lately, I have felt like a private detective in search of the answer to a customer's issue or problem. It's so exciting when I get a strange question, and I'm able to research the issue in my work resources and find the answer. The joy and relief in my voice is felt by the customers, and they can tell I sincerely understood their issue and cared about finding the accurate answer.

Another thing to focus on with understanding and caring about the customers is to detect if they are in a rush. If they are in a hurry (Sometimes they will let you know that they are at lunch or have very little time.), try to ask them what they need you to explain instead of explaining all the details. This was a weak area for me in the past. I used to over-do things by filling the customer with information rather than checking in with them to see what they already knew. If they already know things that are not critical to be covered for their transaction, then let them know that you understand their short time frame, so you will not cover items a, b, or c. Make sure they know what you will not cover, and get their agreement to skip those items if they already have that knowledge.

My manager made me laugh when he told me the following after listening to one of my calls.

"Kokeita, the customer's head will blow off if you give them too much information at one time." The key here is that although we may be full of information on a given topic, we have to focus on the simplest, quickest way to explain the answer to the customer. This saves the customer time and makes the customer service representative available to more customers.

Here are some examples of comments from conference center customers who I spoke with while booking reservations:

"You are infectious! Your chipper personality cheered me up!"

"Wow! You booked my event quickly, and I feel well informed about your facility!"

"I really appreciate the detailed explanation of the fees and guidelines for using this facility."

In addition, you can obtain a better understanding of the customers' concerns by asking open ended questions to determine what type of issue they need you to assist them with resolving. Before asking any questions, always explain to them your reason for asking the question(s) upfront. The way that you ask the questions illustrate that you are concerned and interested in assisting them with finding a solution. It is good to keep in mind that there are three types of questions to ask customers to get a true understanding of their issue. When asking questions to obtain information, you will use these three types: background questions, probing questions, and confirming questions.

A background question is a general question on a particular subject of concern for the customer. Take a look at the example of a background question below.

"Thank you, Ms. Customer, how can I help you today?" (Customer service representative says this.)

"There is a fee on my checking account statement that I don't understand". (Customer says this.)

"Please explain the amount of the fee and what you see on the statement." (Customer service representative says this.)

In the example above, the customer service representative encourages the customer to explain the problem. Next, the representative will have to ask some probing questions to obtain some specific details. The representative will need to ask questions involving some or all of the 5 W's. The 5 W's are always good tools to use (such as who, what, when, why and where). Review the examples of probing questions below.

"I have never seen a fee for copies of checks before." (customer speaking)

"When you get information from the bank, do you usually review it for changes in bank services?" (customer service representative)

"Have you ever ordered copies of checks?"
(customer service representative)

Overall, each question you ask should provide you with the necessary information to diagnose the problem or give you enough information to lead you on an investigative trail to determine the problem and a viable solution. The next line of questions to ask before you start to research the issue are confirming questions. Confirming questions are questions in which you are mainly stating back the specific problem in the form of a question. The goal with confirming questions is to get their agreement to the specific problem(s) before you start researching the issue. There is nothing worse than spending ten or fifteen minutes researching a customer's concern to find out later that the concern you researched is not the issue at all. When you do not ask enough questions to fully understand the issue to research, it is like going to an ice cream parlor to order yogurt and you get ice cream without knowing it. You eat the ice cream and find yourself with unforgettable stomach cramps.

Please see the examples of confirming questions below.

"It appears that the problem is that you have never ordered copies of checks, but there's a check copy fee on your statement. Do I have an accurate idea of your issue?" (customer service representative)

"Yes, please see what you can do to get this corrected!" (customer speaking)

Letting them know that you will need to check a few of your work resources to research their issue makes them feel like you are attaching yourself to their issue and that you are as concerned as they are about getting an accurate answer or the best solution to resolving their issue. I have found that if you sincerely explain that you need to research an issue, they do not mind being put on hold for a few minutes or waiting for a call back. The key here is to "keep your word"! If you do research the issue effectively, and present an accurate answer and a concrete reason for the answer, most customers are able to accept good or bad news. In addition, if you say that you will call the customer back, then call back. When you call them back, you have proven that you and your company do care about their issue and understand their need for an accurate answer or solution.

Below you will find comments from two different customers who spoke with me. These comments illustrate how important it is to show a customer that you care.

"I have talked to a variety of bank employees over the years, but you seem interested in taking care of the customer more than anyone else."

"When I came to this workshop, I was worried about my life outside of the military, but the encouragement and information that I received in your workshop has really inspired me to look forward to the future!"

Overall, the tone and pace of your voice when making comments or asking questions is critical to showing the customer that you understand their concern and wish to do all that you can to help them solve their concern. The golden rule comes in handy here when assisting customers. "Do unto others what you would want them to do unto you". If you always align with the customer and imagine yourself as the customer, it helps you to exhibit the appropriate care and understanding.

CHAPTER THREE
Establishing Credibility and Trust (Building Relationships)

"To know that even one life has breathed easier because you lived. This is to have succeeded.—Ralph Waldo Emerson

We have covered the importance of showing the customers you care about them and understand their issues in addition to the techniques that can be used. In this chapter, you will learn how to establish credibility and trust with every customer. When you establish credibility and trust with your customers, you build relationships. The Merriam Webster Dictionary defines relationship as the act of being connected, interested, and involved. When a customer calls in you want to use all elements of the call to build relationships and create trust.

Every call should include the following four elements: greeting, determination of reason for the call, process transaction and/or provide information, and closing. To establish relationships, the customers must know that you care about them and their issue as well as understand it.

During the greeting, you use this time to exhibit the energy, interest, and competence to handle any issue. Let's examine the reasons why customers call. Most customers call for one or more of these reasons: 1) to process a transaction on their account, 2) to get clarification on a letter that they received from your company, 3) to obtain additional information about your company's products or services, and/or 4) to get help with a problem they are having with their account. During the second step--the determination of reason for the call--you continue to build relationships by keeping these things in mind. While you are trying to determine the reason for the call, always explain why you are asking questions and be sure to state this in a pleasant, calm tone. (Remember to explain everything with good articulation.) While you are listening, remember to employ the CSI technique.

CSI involves:

C – Conclusions **S** – Sensitivity **I** – Interruptions

While the customer is explaining the problem, avoid jumping to conclusions, being overly sensitive, and interrupting the customer. When you jump to conclusions without asking the appropriate probing questions, you end up researching an incorrect issue and waste both your and the customer's time.

When you spend time researching the correct issue, you build credibility with the customers. The customers begin to believe that you are competent with handling their issues. To further build credibility, avoid being overly sensitive to their negative comments about you or the company. It is important to view the negative comments as just the customers' way to blow off steam and cope with their fears and frustration over their issues.

When most people are frustrated about something, they would rather show anger than sadness because anger is viewed as a stronger, more in control behavior while most people view sadness as a weak, out of control type of behavior. If you take the customers' point of view, then you will understand that they don't wish to call in and cry about an account problem that has taken months to correct, so they choose to call in and vent their anger and frustration on you – the customer service representative.

This is a great time to align with their feelings by letting them know that you can understand their anger and assure them that you will do everything you can to find a solution to their problem. You can further build trust by letting them know that you are sorry about the fact that the problem continues to linger. Next, you can strengthen your credibility by telling them your name and contact phone number in addition to assuring them that you will be their point of contact to get the issue resolved.

People like knowing that someone wants to take ownership for a problem. To further establish trust, be sure to check in with them periodically (every 3-5 days) via phone call to give them an update on their issue. This restores their confidence in you and your company, and it also leads to customer retention!

Lastly, be sure to avoid interrupting the customer while you are asking the probing questions. Every time you interrupt customers (especially when they are angry), they just get more upset and feel like your company does not care about them nor their issue. Good listening skills go hand in hand with the appropriate tone. When the appropriate tone is used with good listening skills, you can calm irate customers and build lasting relationships with them.

After you have asked all of the appropriate background, probing, and confirming questions, you can let them know that you will put them on hold to research the issue. When you find the answer to their issue, you need to take a deep breath and present the information in a pleasant, positive, caring tone whether it is good or bad news for the customer. The way that you present the information helps the customer to accept it better. Also, be ready to offer some alternatives to the issue if you cannot give them exactly what they want.

The example below shows how you can turn a potentially negative situation into a positive one.

"When I had my dance in your facility on Friday night the air conditioning system did not work properly. This really put a damper on my event. This may be the last time that I will come to this facility!" (customer)

"Sir, I'm so sorry that this happened during your event, we will immediately refund 20% of the cost of your event. We value your business and want you to continue having your dances here." (customer service representative)

"Great! We will reconsider having events at your facility!" (customer)

Another effective way to deal with irate customers is to be silent and allow them a couple of minutes to vent their frustration. When you do begin speaking, keep your tone pleasant and calm. Your calm attitude helps them to calm down as well.

The fourth and final element of a call is the closing. In the closing, you want to continue building relationships by showing the customers you appreciate them. The first and easiest way is to thank them for calling at the close of the call.

The closing should sound sincere, and it always helps to tell them that you appreciate their business. You can also show them that you appreciate them by listening for opportunities during the conversation to thank them for something. For example, if a customer asks a question about a letter your company sent them, thank them for reading the letter and for calling with the question. Instead of thinking of their call as a bother or unnecessary since they have a letter in front of them, change the way you view the call about the letter. First, think positively about the call and say something like, "Thanks for reviewing the letter and for calling to be sure you understand the letter". There are many different ways to say this, and here are a few examples.

- *Thanks for calling and letting us know you received our letter. I'll be glad to go over it with you.*

- *That's a good question. Thanks for asking. Here's what the letter means by..*

- *I can see how that part of the letter can be confusing. Thanks for calling and asking for another explanation.*

When addressing any customer inquiry or concern, it is helpful to try to see things from the customer's perspective instead of your own.

This requires you to look at the BIG picture and get outside of your worldview. We have a variety of people in this world with a variety of needs and ways to express those needs, so we as customer service professionals have to keep one thing in mind at all times. They call us because they believe we have the answers, and they call us because they need us. Well guess what! We do have the answers, but most of all, we need them too. If they stop calling, we may not be needed for long in our current jobs. Also, we all get our chance to be customers, so let's try to provide them with "outstanding" customer service and we may end up getting it back. I firmly believe what goes around comes back around to us eventually good or bad. Therefore, be patient with the customer. Someone will have to be patient with you soon. ☺

In a world where it appears that everyone is busy and interested in their own destiny and not the needs of others, it is refreshing for customers to talk to someone who has a sincere concern for their issue.

CHAPTER FOUR
Staying Positive

When you work in a call center and take 50 to 100 calls per day, it can be tiring, and you may wonder how to stay positive with so many frustrated people calling. You can stay positive, but it will take a great deal of work on your part at first. The biggest challenges may be changing your view of the customer and learning how to take better care of yourself and your life. You can alleviate the natural stress that comes with life through exercise before or after work or a short brisk walk at lunch time. You can also use positive music before work to get you in the right frame of mind for helping others.

In addition, you can remain positive by getting enough rest each night and eating healthy foods, like lots of vegetables and fruit. These types of foods give you natural energy while snack foods give you quick energy that does not last long. During the day, you are on the phone a great deal, stand up during calls instead of sitting all day. When you do sit, use good posture and sit up straight. To relax tight shoulder muscles, you can roll your shoulders forward and backward and roll your neck around to the left then around to the right about 10 times each.

Also, spend your break or lunch time visualizing yourself in your favorite place (at the beach, mountains, etc.). Lastly, to remain positive, remember not to take things personally and remember the job you have is more than a job; it is a career. Good customer service skills are transferable and are needed in a variety of work environments.

My approach to life and helping others whether at work or at play is pretty much summed up in the quote below by Charles Swindoll.

Attitude

The longer I live, the more I realize
the impact of attitude on life.
Attitude, to me, is more important than facts.
It is more important than the past,
than education, than money, than circumstances,
Than failures, than success,
than what other people think or say or do.
It is more important than appearance,
giftedness or skill.
It will make or break a company,
a church, a home.
The remarkable thing is we have a choice
everyday regarding the attitude we will
embrace for that day.
We cannot change our past . . .
We cannot change the fact that people will
act a certain way.

We cannot change the inevitable.
The only thing we can do is play on the one
string we have, and that is our attitude . . .
I am convinced that life is 10% of what
happens to me and 90% of how I react to it.
And so it is with you . . .

Bibliography

Anderson, Kristin. <u>Great Customer Service On the Telephone</u>. New York: American Management Association. 1992.

Mish, Frederick C. <u>The Merriam-Webster Dictionary</u>. Massachusetts: Merriam-Webster, Incorporated. 2004.

Pickett, Joseph P. <u>The American Heritage College Dictionary</u>. New York: Houghton Mifflin Company. 2002

www.ingramcontent.com/pod-product-compliance
Lightning Source LLC
Chambersburg PA
CBHW061043050326
40689CB00012B/2948